SUPER SOY!

BARBARA FARR

SUPER SOY!

Delicious protein
without meat

KEATS PUBLISHING, INC. New Canaan, Connecticut

Super Soy!

Parts of this book were previously published in *Bestways* magazine

ISBN: 87983-102-2
Library of Congress Catalog Card Number: 76-2982

Published by Keats Publishing, Inc.
36 Grove Street, New Canaan, Connecticut 06840

Printed in the United States of America

This book is dedicated to
the most inspiring health-seeker:
GAYELORD HAUSER

Acknowledgments

The author wishes to express special thanks to *Bestways* magazine for the use of recipes which have previously appeared in "The Healthy Gourmet" column.

Thanks are also due to those friends and co-workers who so patiently tasted all these recipes.

Recipes on pages 43, 125 and 145 are courtesy of Fearn Soya Foods.

Contents

One

THE MAGIC BEAN

Once upon a time—quite recently, in fact—a boy named Jack lived with his mother in a small city. They weren't very rich, but they weren't very poor, either. They were "barely making it". Their clothes were mended, but neat. Their house was snug, sound and mostly paid for. They could have been comfortable, but for one thing—food prices. Every week when Jack took his money to the market, he'd come home with less food. One day, as Jack was on his way to the market, he met a friend who told him about a magic bean. When Jack's mother saw the full grocery sacks he carried home, she was delighted.

"How clever you are," she cried, but soon she became angry and alarmed. "Have you lost your mind, Jack?" the woman shrieked. "There's nothing here but a lot of stupid beans!"

"Trust me," said Jack. "If you don't like the dinner I'm going to make, I'll take you out for a cardboardburger or some ossified chicken." So his mollified mother waited while Jack cooked. The dinner was delicious. Jack's mother was delighted. Jack pointed out that their dinner contained *more* protein and vitamins than their usual carnivorous fare, and that there was no cholesterol in these magic beans, nor had they been shot full of hormones to fatten them for market. After dinner, Jack told his mother the whole story about the magic bean.

The magic bean is called soya or soybean (botanical name: *glycine max*). Although legend claims that the soybean was cultivated in China as early as 2383 B.C., it is certain that it was known by the tenth or eleventh century B.C. Since ancient times, then, the soybean has been the major source of protein in Asia. In the United States, the soybean has been of some economic importance since the 1920s. It has been mainly grown for its oil content and as animal feed, although it is also a component of paints, synthetic rubbers, inks and soaps. Soybeans account for more than 5 percent of all United States exports. They are the biggest single item of United States agricultural export.

In the domestic market, soybean oil is used in 75 percent of our margarine production. Also used in salad oils and shortenings, soybeans are the world's largest single source of edible fat, accounting for about 15 percent of all world supplies. Since soybeans are only about 18 percent fat, the remaining 82 percent is also of great economic importance. As soybean cake and meal, it supplies nearly half of the commercial animal feedstuffs in the United States and is an important export commodity.

Because the soybean is rich in protein, it would be possible, on a balanced vegetarian diet, to feed the entire population of the United States from the soybean production of Kansas, Nebraska, and the Dakotas *alone*. Theoretically, efficient world agriculture could support a global vegetarian population of about 16 billion people. Yet, 10 percent of the world's population is in danger of starving.

The causes of this potential disaster are extremely complex. The poorest, most over-populated countries are those most in need of sophisticated, high-technology farming. Their needs are urgent, and money must be spent to buy food on the world market, rather than farming equipment, although high-technology farming can increase grain yields to almost *six times* that of traditional methods. Distribution of wealth is another factor. Fertilizer shortages in India are causing devastating grain production decreases. Americans use more fertilizer on lawns, golf courses and parks than is used in all of India. In a world where food production and distribution are at best precarious, natural disasters such as floods and droughts have global repercussions.

One of the most obvious factors in the inequality of food availability is the direct correlation between affluence and meat consumption. The rising per capita income in the United States, Europe and Japan is in direct proportion to the rising demand for meat. United States beef consumption doubled between 1940 and 1974. The *average* United States consumption of beef in 1974 was 116 pounds. That means, that on an average, every man, woman and child in the United States derived 30.7 grams of protein per day from beef *alone.* Thirty and seven-tenths grams of protein is equal to about 75 percent of the total daily protein requirement for a 160-pound adult male! We also ate a lot of pork, lamb, poultry, fish, dairy products, eggs, seeds, nuts and grains.

Nutritionally, the over-consumption of protein is a little hard on the kidneys, but otherwise, the body simply converts excess protein to carbohydrate and stores it as fat. Ecologically, the over-consumption of protein is a direct cause of world hunger. The proteinaceous feed now consumed by livestock could be directly consumed by humans. The scenario is this:

1 acre of soybeans provides enough protein to sustain a moderately active man for about 2,200 days.
1 acre of wheat—875 days
1 acre of corn—350 days
1 acre of beef cattle—75 days

One calorie of meat requires four to seven calories of ingested food. In other words, calories and protein desperately needed by hungry people are being fed to livestock!

Water is another item of global importance. Water is used in all phases of food production from growth to table. One must even consider water used to wash or cool processing equipment, etc. Based on all these factors, water consumption on a mixed animal and vegetable diet is estimated at 2,500 gallons *per person per day,* while an all-vegetable diet consumes about 300 gallons. A pound of meat requires about 25 times as much water as a pound of vegetables. Whether one becomes a full-time vegetarian or not is a matter of personal conviction, but it is in the interest of every person now living, as well as future generations, to give up meat in at least some of our meals.

If a "meatless meal" sounds like deprivation of some sort, think of good hot cereal breakfasts, cool salad lunches, hearty bean soups with good whole wheat bread. Meals in which grains, nuts, seeds and legumes replace meat are some of the best in the world. All kinds of delicious dishes can be made without meat, so reducing some percentage of your meat consumption can be a real pleasure—an adventure in good nutrition, creative cooking and good taste.

Soybeans are one of the most versatile foods for meatless cooking. Some very interesting foods based on soybeans have been developed in Asia, where soybeans are often the only source of high-quality protein. Some of these foods are known in the United States, others are becoming more widely known and available. Perhaps as more people learn to appreciate these foods, and depend on them as protein sources, some of the lesser-known varieties will become popular. Tofu (see index) can be bought in many different forms, shapes and textures. If you find that you like tofu and you are lucky enough to have access to a Japanese grocery, try some varieties. In this book, we have concentrated on those forms of soybeans which are most available. Health food stores usually carry the widest variety of soy products.

Two

SOYBEANS AND NUTRITION

Nearly all of us learned in school that there are four generally accepted food groups. These are the meat group, the vegetable-fruit group, the milk group, and the bread-cereal group. We were taught that we should eat several foods from each group every day. While it is true that the growing infant definitely needs milk every day—mother's milk—it is also true that grains, legumes and cereals can adequately supply most of the needs provided by the meat and milk groups. Vitamin B-12 must be supplemented in an all-vegetable diet, and sources of calcium, minerals and other vitamins should be considered.

Another meaningful grouping of foods would consist of three groups: protein, carbohydrate and fats. These groups overlap, many foods containing all three. The body must ingest certain fats and oils; these are "essential fatty acids." Carbohydrate is broken down into simple sugars, one of which, glucose, is the brain's *only* "fuel." Protein is the actual matter of which body cells are formed. Proteins are also responsible for various functions which take place within the body. Certain proteins are manufactured in the body; others must be ingested. There are eight amino acids (proteins) which *must* be ingested and are therefore considered "essential." These eight are tryptophan, threonine, phenylalanine, lysine, valine, isoleucine, methionine-cystine and leucine. They function in a specific ratio, so that if a food is deficient in any one of the eight, the value of the remaining seven is lowered proportionately.

This deficiency must be remedied immediately. You cannot eat a food deficient in lysine at lunch and "correct" the deficiency two hours later. This corrective balancing of the eight essential amino acids is "protein complementarity." For example, soybeans are somewhat low in methionine-cystine. Cereals and grains are rich in methionine-cystine but low in lysine, which is abundant in soybeans. Thus, soybeans plus cereals and grains in the same meal provide more protein than either could if eaten separately. In planning meals based on soybeans, remember to include wheat, sesame, peanuts, corn, brown rice, eggs or milk products.

Recent FDA (Food and Drug Administration) recommended daily allowances call for 45 to 65 grams of protein daily for an adult. (Needs of growing children, pregnant or lactating women differ.) One can attain that much protein in a pound of lean steak, or 8 eggs or a quart of milk. Soybean granules provide almost 50 grams of protein in 100 grams while steak provides 20 to 30 grams per 100 grams. Soybean curd is equivalent in calcium to cow's milk.

Certain minerals, such as iodine and zinc, should be supplemented on a vegetarian diet. Iodine occurs in sea vegetables such as kelp and nori, or one can use iodized salt or sea salt. Zinc is present in vegetables grown on zinc-rich soil. We were once subject to a zinc deficiency when we moved from one state to another that had zinc-poor soil, on which we grew our vegetables. We corrected it by both dietary supplements and adding zinc to the soil.

Vitamin B-12 is manufactured by animals, but the human body seems unable to manufacture its own. Unless eggs and dairy products are included in a vegetarian diet, B-12 must be supplemented.

Another factor in the diet which is becoming recognized as increasingly important is dietary fiber, which is basically indigestible carbohydrate. Soybeans, legumes, whole grains, nuts, seeds and vegetables all contain significant amounts of fiber. Meats, poultry, seafood, eggs and dairy products contain *none.* The fiber may be finely ground as in flours and powders, for it is an indigestible molecule which gives the bulk, "roughage" and absorbency to food waste traveling through the body.

Our "magic" bean, then, contains fat (essential fatty acids), carbohydrates (including dietary fiber but no starch), and a lot of protein. It also contains calcium, iron, potassium and phosphorus and has an alkaline ash, which makes it valuable in a diet containing acid-forming grains and meats. Soybeans also contain vitamin A and vitamins of the B complex. When the beans are sprouted, they contain all of the above nutrients as well as vitamin C.

Dieting is one of the most popular national pastimes. People who are not actually dieting are usually regretting their last diet, or planning their next one. Thin is healthy, but dieting is boring, at best. Eating is one of the most basic necessities of life, and should be a pleasure. In fact, studies have shown that people who are relaxed and happy while eating have fewer digestive problems than those who are tense and unhappy.

It is possible to diet very happily. The trick is to diet constantly, which really means not dieting at all, just eating rationally. That means getting the necessary amount of vitamins, minerals, fats, carbohydrates, and protein. Fats and carbohydrates are relatively easy to obtain. Vitamins and minerals can be supplemented when not readily available in the foods one eats. Protein, however, is traditionally obtained from meats, fish, and poultry. Vegetarian sources of protein are plentiful, but not as well known or understood as animal protein. Even so, it is possible to get the proper nutrients. The trick is to obtain all of the necessary nutrients *without* getting too many calories. Excess calories are fattening. Period. Even if the excess calories are derived from protein, if they are in excess of the body's need, they will eventually become fatty tissue.

Nearly all foods contain some protein, but it is important to seek out those foods that offer the most protein with the fewest calories. Soybeans, cheese, seeds, eggs, nuts, milk, wheat germ and nutritional yeast are excellent sources of protein. Tofu is also an excellent source, providing about 69 calories and 7.5 grams of protein in a 4-ounce cake. Brewer's yeast contains 4 grams of protein in a 23-calorie tablespoonful. Soy powders and granules can be used in many dishes to boost protein at a relatively low cost in calories.

Your need for calories is variable, depending on your weight, rate of physical activity, etc. Obese and sedentary persons need fewer calories than do those who are active and muscular. It's rather like the caterpillar's mushroom in *Alice in Wonderland:* too many calories will make you fat, too few calories will make you thin. Protein is another matter. The human body tissue is mainly composed of protein, some of which can be synthesized by the body, some of which must be ingested. Gayelord Hauser suggests 1 gram of protein for each 2.2 pounds of ideal body weight. That is for a normal adult. Special needs may increase the need for protein.

It is not too difficult to learn to balance your protein needs with the proper number of calories. The first and easiest step is to eliminate those foods that offer only empty calories, such as sugar and highly-refined foods. Sliced bread, for example, averages about 70 calories per slice. Protein content varies from about 2 grams to about 9 grams per slice. Obviously, the 9-gram slice gives you more for your calories.

WHY WE DO NOT USE TEXTURED VEGETABLE PROTEIN

Textured vegetable protein (TVP) appears in many products, from meat analogs* to extenders for processed meats. Because it is basically soy protein, it has been hailed as a solution to high meat consumption. However, it has many disadvantages which are not as widely touted.

First, most methods for producing TVP require isoelectrically precipitated soy protein. The drawbacks in isoelectric precipitation are that the resulting globulins combine with phytates and other low-molecular-weight compounds which decrease the availability of zinc and other essential minerals (1) and that some of the proteins become insoluble.(2)

The spinning method for producing TVP requires dispersion of the soy protein isolate in a 14 to 18 percent concentrated solution of sodium hydroxide, a pH of 10 to 11 and aging at 40° to 50° C. until the dispersion (called "spinning dope") reaches a spinnable state. It is then forced through a spinnerette into a coagulating bath containing acid and salts. Flavor is "improved" by treatment with sulfur dioxide. (3)

Heat gelation adds the soy protein isolate to hot solution of trisodium phosphate, then mixing it with fat, emulsifier, color, smoke and other flavorings. (4) Another method calls for granular particles of the isolate to be soaked in water containing calcium hydroxide or trisodium phosphate. (5)

The extrusion method uses soy flour. (About 80 percent of the soy flour produced in this country is processed with hexane.) The flour is mixed with fat, flavorings, carbohydrates, colorings and sulfur or sulfur compounds. The mixture is then passed through an extruder under set conditions of moisture, temperature, pressure and time. The resulting mixture expands as it is extruded, and is cut to size by revolving knives. (6) TVP is a chemical feast indeed!

References

1 Smith and Circle. *Soybeans: Chemistry and Technology, Vol. 1.* Westport, Conn.: AVI Publishing, 1972. pp. 108, 185.
2 Ibid., p. 100.
3 Ibid., p. 365 ff.
4 Ibid.
5 Ibid.
6 Ibid.

* Analog: something that is analogous or similar to something else.

Three

A GLOSSARY OF SOY (AND OTHER THINGS)

Since some of the products used in this book may be unfamiliar, we are presenting a glossary with a bit of information on each one. We have also attempted to explain some of the questions that arise, such as the difference between soy "powder" and soy "flour."

We also include in our glossary some items that may be unfamiliar but are very useful in soybean cookery. You will find that tahini, for instance, is not only a marvelous ingredient in many dishes, but is made from sesame seeds and is a complement to soy protein.

Where nutritional information is available, we have included it. This information may often be very general, and varies from one product to another. Many manufacturers have nutritional information on their packaging, which should be accurate as it reflects the results of their particular processing methods.

If you are interested in dietary fiber, it may be estimated by multiplying the amount of crude fiber by 3 or 4.

Bran

Bran is the outer hull of the wheat kernel. Although it contains some vitamins and minerals, it is primarily used for its high fiber content. One way that we find handy to use bran is to whir it in a blender until it becomes a powder. Bran absorbs liquid, so recipes should be adjusted accordingly.

Nutritional Information
(per 100 grams)

Protein	16	g.
Fat	4.6	g.
Calories	213	
Total Carbohydrate	61.9	g.
Crude Fiber	9.1	g.
Calcium	119	mg.
Phosphorus	1,276	mg.
Iron	14.9	mg.
Sodium	9	mg.
Potassium	1,121	mg.
Thiamine	.72	mg.
Riboflavin	.35	mg.
Niacin	21	mg.

Dried soybeans

This is the most versatile, most widely available form of soybeans. Dried soybeans may be stored at room temperature and are simple to cook. (See p. 41.) Dried soybeans contain antinutritive factors which are destroyed by cooking, so that if dried beans are ground at home for soy milk or other recipes, caution should be observed. Home-ground soy powder should only be used in recipes where cooking is required. Many varieties of soybeans are grown; some are more delectable than others. When shopping, look for soybeans labeled "Cooking Soybeans." If you grow your own, ask your local United States Agriculture Extension agent to recommend a variety for superior edibility.

Nutritional Information

(per 100 grams)

	Dry		*Cooked*	
Protein	34.1	g.	11	g.
Fat	17.7	g.	5.7	g.
Calories	403		103	
Total Carbohydrate	33.5	g.	10.8	g.
Crude Fiber	4.9	g.	1.6	g.
Calcium	226	mg.	73	mg.
Phosphorus	554	mg.	179	mg.
Iron	8.4	mg.	2.7	mg.
Sodium	5	mg.	2	mg.
Potassium	1,677	mg.	540	mg.
Vitamin A	80	I.U.	30	I.U.
Thiamine	1.10	mg.	.21	mg.
Riboflavin	.31	mg.	.09	mg.
Niacin	2.2	mg.	.6	mg.

Green soybeans

Green soybeans are not widely available, and are not as versatile as the more mature, dried bean. Canned green soybeans are available in some areas, and fresh green soybeans are a delicious cooked vegetable either alone, or added to soups, stews or casseroles.

Nutritional Information
(per 100 grams)

Protein	9.8	g.
Fat	5.1	g.
Calories	118	
Total Carbohydrate	10.1	g.
Crude Fiber	1.4	g.
Calcium	60	mg.
Phosphorus	191	mg.
Iron	2.5	mg.
Vitamin A	660	I.U.
Thiamine	.31	mg.
Riboflavin	.13	mg.
Niacin	1.2	mg.
Ascorbic Acid	17	mg.

Lecithin

Lecithin is a product of soybean oil, and is present in all natural soy products. An emulsifier, lecithin helps to break up fats and hold them in suspension. It contains cephalin and inositol phosphatides as well as tocopherols which make it a natural antioxidant. Lecithin may be purchased as granules or liquid. Granular lecithin may be sprinkled on cereals and other foods; liquid lecithin is excellent for use in cooking. One special property of lecithin is that it makes a non-greasy, non-stick coating for baking pans and casseroles. It's a little gooey to spread, so we usually put about a teaspoon of vegetable oil in the baking pan, then add about a teaspoon of liquid lecithin and spread them together.

Miso

Miso is a seasoning paste made from a fermentation of soybeans, salt and water. Tasty and nutritious varieties of miso may also contain rice or barley. Miso is available in health food stores, Japanese markets and some supermarkets. Miso soup is a staple in the Japanese diet, where it is often served for breakfast. Miso is traditionally stored at room temperature, but we prefer to keep it refrigerated. We suggest that you try several varieties of miso, as each has its own special flavor.

Nutritional Information
(per 100 grams)

Protein	10.5	g.
Fat	4.6	g.
Calories	171	
Total Carbohydrate	23.5	g.
Crude Fiber	2.3	g.
Calcium	68	mg.
Phosphorus	309	mg.
Iron	1.7	mg.
Sodium	2,950	mg.
Potassium	334	mg.
Vitamin A	40	I.U.
Thiamine	.06	mg.
Riboflavin	.10	mg.
Niacin	.3	mg.

Nutritional yeast

This yeast is especially grown for human consumption and should not be confused with baker's yeast. Nutritional yeast comes in many forms and flavors and it is helpful to shop around until you find one that suits your taste. All nutritional yeasts are high in protein and B vitamins, but the proportions of B vitamins vary widely. Read and compare labels to choose the yeast that is best for you. Due to the difference between brands, the following information is very general.

Nutritional Information
(per 100 grams)

Protein	38.8	g.
Fat	1	g.
Calories	283	
Total Carbohydrate	38.4	g.
Crude Fiber	1.7	g.
Calcium	210	mg.
Phosphorus	1,753	mg.
Iron	17.3	mg.
Sodium	121	mg.
Potassium	1,894	mg.
Thiamine	15.61	mg.
Riboflavin	4.28	mg.
Niacin	37.9	mg.

Sesame tahini

Sesame tahini is a paste made from ground sesame seeds. It is a bit thinner than peanut butter and is bland, as it has no salt added. It can be used as a spread or sauce, and can be mixed with other ingredients for endless variety. The protein in sesame is complementary to that in soy.

Soy granules

Soy granules are also called soy grits. Granules are toasted nuggets of soybeans from which most of the oil has been extracted by pressure, or by chemical solvents. Read labels to be sure that those you are buying have not been processed with solvents. Granules are nutritionally equivalent to low-fat soy powder, but their nut-like flavor and texture open a broad range of uses. Added to breads and muffins, they add a nut-like texture while boosting the protein. They may be used to replace noodles or rice in soups, or as a topping for sweet or savory casseroles. They make an excellent garnish for salads, and may be added to hot or cold cereals. When moistened with water, soy granules become soft and swollen as they readily absorb liquid.

Nutritional Information
(per 100 grams)

Protein	48.28	g.
Fat	6.8	g.
Calories	311	
Total Carbohydrate	33	g.
Crude Fiber	2.5	g.
Calcium	149	mg.
Phosphorus	643	mg.
Iron	8.38	mg.
Sodium	6.05	mg.
Potassium	1,859	mg.
Vitamin A	92	I.U.
Thiamine	1.7	mg.
Riboflavin	.59	mg.
Niacin	1.65	mg.
Ascorbic Acid	9.2	mg.

Soybean oil

Soybean oil contains large amounts of linoleic acid, one of the essential fatty acids. Crude soy oil also has lecithin in it. Dry soybeans and full-fat soy powder contain this valuable oil in significant amounts. Crude soybean oil is not used in cooking because it foams, but blends of vegetable oil suitable for cooking are available in health food stores. Our favorite is a combination of soy, peanut and safflower which contains a good blend of essential fatty acids.

Soybean powder

Soy powder is available in several versions from natural (full-fat) to defatted. All powders should be heat-processed.

> FULL-FAT SOY POWDER: The superior brands are entirely natural, and have been heat-treated to remove any anti-nutritive factors. Full-fat powder contains about 22 percent fat in the form of soy oil, which contains lecithin and tocopherols which act as natural preservatives.
>
> LOW-FAT SOY POWDER contains about 6 percent fat. The fat may be removed either by pressure (natural) or by hexane solvent (chemical). Be sure to read the label before buying any low-fat powder.
>
> DEFATTED SOY POWDER is, by definition, about 1 percent fat. In order to achieve this level, the fat is usually removed by the use of a petroleum-based solvent. Defatted soy powder is not widely available.

NOTE: The terms "soybean powder" and "soybean flour" are sometimes used interchangeably. Both refer to *particle* size, i.e., finely-ground soy. However, we use the term "powder" because it refers to a product which is not only finely ground but heat-treated as well, so that the anti-nutritive factors present in raw soybeans have been destroyed. Also, "flour" usually refers to a powdered grain in which gluten is present. Since the action of glutinous flours is very different from that of non-glutinous "powders," we believe that the term "powder" is less confusing. For instance, although the addition of soy powder will boost the protein in any recipe calling for flour,* soy does *not* contain any gluten and is *not* a substitute for flour.

*2 tbs. soy powder added to a cup of *un*enriched white flour *doubles* thiamine, calcium and iron and adds 50 percent more protein!

Nutritional Information

(per 100 grams)

	Full-fat soy powder	*Defatted soy powder*
Protein	40.26 g.	47 g.
Fat	22.03 g.	.9 g.
Calories	418	326
Total Carbohydrate	26 g.	38.1 g.
Crude Fiber	2.4 g.	2.3 g.
Calcium	138 mg.	265 mg.
Phosphorus	558 mg.	655 mg.
Iron	5.8 mg.	11.1 mg.
Sodium	7.26 mg.	1 mg.
Potassium	1,660 mg.	1,820 mg.
Vitamin A	137 I.U.	40 I.U.
Thiamine	1.98 mg.	1.09 mg.
Riboflavin	.44 mg.	.34 mg.
Niacin	1.4 mg.	2.6 mg.
Ascorbic Acid	7.8 mg.	0

NOTE: Low-fat soy powder is nutritionally equivalent to soy granules.

Soy protein isolate

As the name implies, soy isolate is the protein fraction of soybeans which remains after nearly all of the other components have been removed. Isolate is at least 90 percent protein on a dry basis. It is used as a protein supplement, and is best used with milk or sesame tahini to add complementary amino acids. It is most often used in blender or milk-shake-type drinks.

Soy sauce

Japanese soy sauce is called "shoyu." It is a seasoning liquid made from a fermented mixture of soybeans, salt and wheat. It differs from Chinese soy sauce, which is not made with wheat. All recipes in this book have been made with Japanese shoyu. Some products marketed as soy sauce are actually mixtures of hydrolyzed vegetable protein with caramel coloring and flavoring. Good Japanese shoyu may be labeled "tamari."* Always read the label carefully; the *only* ingredients in shoyu are soybeans, salt, wheat and water. Shoyu keeps well at room temperature but, for the best quality, we prefer to keep on hand only as much as will be used within a month.

*"Tamari" is the name of an ancient Japanese seasoning which preceded shoyu. Tamari refers to a liquid residue from miso production, but some shoyu marketed in the United States is labeled "tamari." To avoid further confusion, we urge you to read ingredient statements carefully.

Nutritional Information
(per 100 grams)

Protein	5.6	g.
Fat	1.3	g.
Calories	68	
Total Carbohydrate	9.5	g.
Crude Fiber	0	
Calcium	82	mg.
Phosphorus	104	mg.
Iron	4.8	mg.
Sodium	7,325	mg.
Potassium	366	mg.
Thiamine	.02	mg.
Riboflavin	.25	mg.
Niacin	.4	mg.

Soybean sprouts

Dried soybeans are seeds, and will germinate under proper conditions. Special varieties of soybeans are grown for their sproutability and are marketed in health food stores as "Sprouting Soybeans." (For sprouting directions, see p. 42.) The "bean sprouts" used in oriental cooking are usually mung bean sprouts which are easier to grow and more palatable eaten raw. Soybean sprouts generate vitamin C as they grow, but have a distinctive raw bean flavor and are usually cooked slightly before inclusion in salads and other dishes. According to the United States Department of Agriculture, cooked, *drained* soy sprouts contain about 60 percent less vitamin C than raw sprouts. We suggest cooking the sprouts as little as possible (see p. 48) and *using* the cooking water. Sprouted soybeans are sometimes available in Chinese markets.

Nutritional Information
(per 100 grams of raw sprouts)

Protein	6.2	g.
Fat	1.4	g.
Calories	46	
Total Carbohydrate	4.3	g.
Crude Fiber	.8	g.
Calcium	48	mg.
Phosphorus	67	mg.
Iron	1	mg.
Vitamin A	80	I.U.
Thiamine	.23	mg.
Riboflavin	.20	mg.
Niacin	.8	mg.
Ascorbic Acid	13	mg.

Tofu

Tofu is a bland, easily digestible "cheese" made from a coagulation of soy milk. Home versions may be made using lemon juice or vinegar as a coagulant (see p. 43). Commercial tofu is coagulated with *nigari* (either calcium chloride or magnesium chloride) or with magnesium or calcium sulfate. Many very interesting variations of tofu are available in Chinese and Japanese groceries, but fresh tofu must be purchased freshly made or made at home. An "instant" variety is available which includes a packet of soy powder with a smaller packet of coagulant. Due to the bland taste, tofu may be included in a wide variety of dishes, and lends itself to all sorts of seasonings. Tofu is also called dowfu (Chinese) and bean curd.

Nutritional Information
(per 100 grams)

Protein	7.8	g.
Fat	4.2	g.
Calories	72	
Total Carbohydrate	2.4	g.
Crude Fiber	.1	g.
Calcium	128	mg.
Phosphorus	126	mg.
Iron	1.9	mg.
Sodium	7	mg.
Potassium	42	mg.
Thiamine	.06	mg.
Riboflavin	.03	mg.
Niacin	.1	mg.

Yogurt

Yogurt is formed by the action of bacteria at controlled temperatures. Both soy milk and cow's milk may be used to make yogurt. (See p. 46.) We prefer yogurt to sour cream in uncooked recipes. Stirring yogurt liquifies it somewhat, so if you want it thick, avoid stirring as much as possible. Fresh yogurt may be kept refrigerated for about a week. We like its tangy flavor, and it is very good with chopped fresh vegetables and herbs, or chopped fruit and nuts.

Nutritional Information

(per 100 grams of skim-milk dairy yogurt)

Protein	3.4	g.
Fat	1.7	g.
Calories	50	
Total Carbohydrate	5.2	g.
Crude Fiber	0	
Calcium	120	mg.
Phosphorus	94	mg.
Sodium	51	mg.
Potassium	143	mg.
Vitamin A	70	I.U.
Thiamine	.04	mg.
Riboflavin	.18	mg.
Niacin	.1	mg.
Ascorbic Acid	1	mg.

Four

BASIC TECHNIQUES WITH SOYBEANS

Cooking soybeans

Method 1

Soak the soybeans 6 hours or overnight. 1/3 cup dry = 1 cup soaked. Freeze the beans in soaking water, then cook for about 1 hour.

Method 2

Cook the soaked soybeans in a pressure cooker at 15 lbs. (standard) pressure for 45 minutes.

Method 3

Cook the soaked soybeans over low heat for several hours. Add water as needed. Skim foam from the beans when they reach a boil.

Sprouting soybeans

Soybeans are notoriously difficult to sprout. Be careful to purchase beans that are labeled "for sprouting," as there are some varieties that are a bit easier to sprout.

Cover soybeans with 4 times their volume in water (2 cups of water to 1/2 cup of beans). Water should be about room temperature. Let stand about 8 hours (a bit longer in a cool room). Drain the soaking water and use it for cooking. Place the beans in a colander and rinse well with cool water.

Place the beans in a commercial sprouter, or in a large jar. Cover the mouth of the jar with wire mesh screening, cheesecloth or nylon net and secure with a rubber band or string. Tip the jar, mouth downward, so that it can drain. Keep sprouts in a dark place at cool room temperature.

Sprouts should be rinsed 3 times daily until maturity. To rinse: Gently place the beans in a colander (the sprouts are fragile; don't break them). Rinse with cool water, remove any broken, soft or fuzzy beans, and rinse again. Drain, and return the beans to whatever sprouting device you are using. If you use a jar with a screened top, you can pour the water through the top without removing the beans, drain the jar upside down, and return it to its tipped position to drain thoroughly. Sprouting seeds like a temperature of about 70° to 80° F.

Harvest sprouts when they are ½-inch to 2-inches long (4 to 6 days). ½ cup of dry seeds yields about 4 cups of sprouts.

Tofu (soya bean curd)

4 *cups water*
1 *cup soya powder*
4 *tbs. lemon juice*

Mix the soya powder and water together and let stand 20 minutes, stirring occasionally.

Bring the mixture to a boil in a very large pot. Stir constantly. Reduce heat and simmer 5 minutes. Remove pan from heat and add lemon juice all at once, while stirring.

Allow mixture to cool to room temperature. Drain mixture into a cheesecloth-lined colander. Wrap the drained curds in the cheesecloth and hang to drain for about an hour. Gently remove the cloth.

Whole or sliced tofu may be stored in the refrigerator, covered with water, for several days.

Soy milk

4 cups water
1 cup soy powder

Mix water and soy powder; let stand at room temperature for 2 hours.

Bring to a boil in a very large pot (it will foam up to about triple its original volume), stirring constantly. Let cool. If you wish, you may strain the soy milk through cheesecloth.

Refrigerate. Shake well before using.

The above recipe is used for drinking and for yogurt. For drinking, you may flavor to taste with honey, molasses, fruit syrups, etc.

SOY MILK FOR COOKING

The above recipe may be used in all recipes calling for soy milk, but for a quick substitute, you may simply mix soy powder and water together. The proportions are 4 parts water to 1 part soy powder. Thus, if a recipe calls for 1 cup of soy milk, you may mix 1 cup of water with 1/4 cup of soy powder.

Soy cheese

1 *cup low-fat soy powder*
2 *cups water*

Mix the soy powder into the water until smooth. Place in a glass bowl and let stand at room temperature for 3 to 4 days. Cheese should have a fresh, slightly sour smell.

Drain the cheese into a cheesecloth-lined sieve.

Wrap the cheese in the cheesecloth and hang to drain for 2 to 3 hours. Season to taste.

Soy yogurt

3 cups soy milk

1 cup yogurt

Heat the soy milk to 120° F. Stir in the yogurt, and mix very thoroughly. Incubate in a yogurt maker until set, or place in a covered glass baking dish and incubate in a gas oven overnight. (The pilot light usually provides enough heat to incubate the yogurt.)

These directions call for the use of commercial yogurt as an ingredient. Several brands will be available at any good health food store.

You may also use a yogurt culture, following the directions on the package.

Soy butter

1/4 cup soy oil
3/4 cup soy powder

Mix together. Season to taste with vegetable salt, herbs, spices, etc. Chopped nuts or seeds may be added.

Steamed soy sprouts

1/4 cup water
salt
2 cups soy sprouts

Heat the water to boiling and add a pinch of salt. Add the sprouts. Cover the pot tightly and steam for about 4 minutes.

Soy béchamel sauce

2 tbs. oil
4 tbs. soy powder
1 cup stock or water
2 tsp. liquid lecithin

In a large saucepan, heat the oil. Add the soy powder, stirring with a wire whisk to mix well. Add the stock, stirring constantly with the whisk.

Bring to a boil (it will foam), reduce the heat and simmer for about 5 minutes, stirring often.

Add the liquid lecithin and stir with a whisk until the sauce thickens.

Makes about 1 cup.

This is a basic recipe and contains no seasoning. You may add salt, seasoned salt, soy sauce, pepper and/or herbs to taste. Sautéed onions or mushrooms may also be added. Use the béchamel sauce to make creamed vegetables, or use as a sauce or gravy over sliced bean loaf, etc.

Soy mayonnaise

1/2 cup water
1/2 cup soy powder
1/2 tsp. salt
1/2 tsp. dry mustard
1/4 tsp. garlic powder (optional)
1/4 cup fresh lemon juice
3/4 cup vegetable oil
1/4 cup olive oil

Place the water, soy powder, salt, mustard, garlic, lemon juice and 1/4 cup of the oil in a blender and blend until well mixed. With the blender on low speed, trickle in the remaining oil until all oil is blended in and the mayonnaise is thick. Chill.

Mayonnaise may be varied by the addition of chopped fresh herbs or parsley.

Makes about 1 cup.

Five

APPETIZERS AND SNACKS

Cocktail cheese ball

3/4 cup soy cheese (see p. 45)
2 tsp. caraway seed
1/2 tsp. paprika
1/4 tsp. turmeric
1/8 tsp. powdered mustard
sea salt to taste
4 tbs. tahini

Mash the cheese. Add the caraway seed, spices and salt. Mix in the tahini. Form into a ball. Chill before serving. Vary the seasonings to suit yourself. The tahini and salt are the only essential ingredients. Try chopped nuts, herbs, etc.

Guacamole

1 avocado
1 tbs. lemon juice
2 tbs. chopped onion
1 small clove garlic, minced
2 tbs. low-fat soy powder
1/2 tsp. chili powder
sea salt to taste

In a blender, whiz avocado, lemon juice, onion and garlic until fairly smooth. Remove to a bowl and mix in the remaining ingredients.

Serve chilled as a dip, spread or salad garnish.

Makes about 1 cup (depending on the size of the avocado).

Artichoke hors d'oeuvres

1 8½ oz. can artichoke hearts

1 egg

2 tbs. oil

1/3 cup whole wheat bread crumbs

3 tbs. soy powder

1 tsp. oregano

1 tsp. basil

2 tbs. grated Parmesan cheese

1/2 tsp. garlic powder

1 tsp. sea salt

1/2 tsp. pepper

In a large bowl, mash the artichoke hearts and stir in the egg. Heat the oil in a large skillet. Add all the remaining ingredients, and stir over moderate heat for 2 to 3 minutes. Let cool. Mix thoroughly into the artichoke mixture, and chill for 1 hour.

Heat oven to 300° F.

Form artichoke mixture into small balls. Roll in extra bread crumbs, and bake on ungreased cooky sheet for 15 minutes. Serve warm. Makes about 2 dozen 1½-inch balls.

Soy felafels

1½ cups (about 2 slices) crumbled fresh whole wheat bread
1/2 cup water
2 cups cooked, drained soybeans
1/4 cup lemon juice
2 cloves garlic, crushed
1 tbs. chopped parsley
1/2 tsp. cayenne pepper (optional)
1/2 tsp. cumin powder
1/2 tsp. basil
1/2 tsp. sea salt
black pepper
2 eggs, lightly beaten
dry whole wheat bread crumbs
vegetable oil

Place the fresh crumbs in a bowl and pour the water over them. Let stand 10 minutes.

Grind the soybeans, or make a rough puree in the blender, adding the lemon juice.

Squeeze as much water as possible from the bread crumbs. Mix together the crumbs, beans with lemon juice, garlic, spices and eggs. Mix thoroughly. Add as many dry crumbs as needed to make the mixture hold its shape lightly.

Heat as much oil as needed to deep-fry the felafels. Heat to 375° F. Form felafel mixture into 1-inch balls and fry until golden brown. Serve hot.

Makes about 3 dozen 1-inch balls.

Mushrooms without caviar

18 raw, medium-sized mushroom caps
1 tbs. oil
1 tbs. minced onion
3 tbs. soy granules
1 tbs. nutritional yeast
1 tsp. tomato paste
2 tbs. soy sauce
1/4 cup water

Wash and dry the mushroom caps and place them in a shallow, oiled pan.

Heat the oil in a skillet and sauté the onion until limp.

Add soy granules and the remaining ingredients. Stir until well mixed and all liquid is absorbed.

Place about 1/2 tsp. of the granule mixture in each mushroom cap. Broil the stuffed mushroom about 3 inches from the flame for about 3 minutes or until quite brown.

Near-Eastern tahini spread

1 cup cooked soybeans
1/3 cup tahini
1-2 cloves garlic, crushed
2 tbs. lemon juice
dash cayenne pepper
1/2 tsp. sea salt
fresh parsley, chopped

Put all ingredients except parsley in a blender and blend until smooth.

Pile in a bowl and garnish with lots of chopped parsley. Serve as a spread for thin crackers, or as a dip.

Dip Japonaise

2 scallions

2 tbs. miso

1 tbs. tahini

1/2 tsp. fresh grated ginger (optional)

2/3 cup soy yogurt (or plain dairy yogurt)

Mince the scallions, keeping the green and white parts separate.

Mix the miso, tahini, ginger and white part of scallion. Stir in the yogurt. Garnish with the green part of the scallion.

Chill and serve as a dip with fresh vegetables, or heat in a double boiler and serve as a warm dip or sauce for cooked vegetables.

Soy snacks

Soak soybeans overnight in salted water. Boil one hour.

Heat oven to 350° F. Spread the beans on a baking sheet to form a single layer. Bake 30 minutes, stirring every 10 minutes or so. If desired, add salt or seasoning while hot.

Store in a tightly covered jar.

Granule tabbouleh

1/2 cup soy granules
1/2 cup bulgur wheat
2 cups water
1/3 cup lemon juice
1 cup chopped parsley
1 medium onion, minced
2 large tomatoes, minced
1/3 cup oil (preferably olive oil)
1 cup chopped fresh mint (or 1 tbs. dry)
sea salt and pepper to taste
1/4 cup soy granules

Soak the 1/2 cup of soy granules and the bulgur wheat in the water for 10 minutes.

Line a sieve with cheesecloth and drain the grains. Wrap them in the cloth and sqeeze to remove as much water as possible. Toss with the lemon juice and chill thoroughly.

To serve, toss with the remaining ingredients.

Serves 6.

Soybean salad

2 cups cooked soybeans
1/4 cup vinegar
2 tbs. soy sauce
1 tbs. molasses
1 small carrot, chopped
2 tbs. minced sweet pepper
2 tbs. minced sweet onion
1 tbs. minced parsley

Combine soybeans with vinegar, soy sauce and molasses. Bring to a boil, cover and simmer 10 minutes. Let stand until cool. Chill. Combine with remaining ingredients.

Serves 4.

Soybean salad 2

2 cups cooked soybeans
1/4 cup sweet relish
1/4 cup chopped, pitted black olives
2 tbs. diced onion
2 tbs. minced celery
2 tbs. minced green pepper
2 hard cooked eggs, diced
1/4 cup wine or cider vinegar
sea salt and pepper
1/4 cup olive oil
1/2 cup vegetable oil
lettuce

Combine soybeans, relish, olives, onion, celery, pepper and eggs in a large glass or china bowl.

Place the vinegar in a smaller bowl. Add salt and pepper. Combine the oils and stir into the vinegar with a fork until well mixed. Pour over the soybean mixture and toss lightly. Chill well.

Arrange lettuce on plate. Remove soybean salad with a slotted spoon and mound on lettuce.

Serves 6.

Granule sandwich spread

1/2 cup soy granules
3/4 cup water
1/4 cup soy sauce
4 tbs. mayonnaise
1 tbs. tahini
1/4 cup soy granules

Cook the 1/2 cup soy granules with the water and soy sauce until nearly all of the liquid is absorbed. Let cool. Stir in the remaining ingredients. Add chopped vegetables and herbs to taste.

Makes about 1½ cups.

Six

SOUPS AND BREAKFAST DISHES

Bean sprout soup with miso

*1 piece kombu seaweed, about 6 inches by 3 inches**
4 cups water
1/4 tsp. grated fresh ginger
1 small onion, minced
1 medium carrot, thinly sliced
1/4 cup fresh or frozen peas
1 small potato, diced small
4 tbs. miso
1 cup bean sprouts
1/2 cup diced tofu (optional)

Place kombu and water in a pot and bring to a boil. Add the ginger and let simmer 3 minutes. Remove kombu and discard. Add the onion, carrot, peas and potato. Return to the boil and simmer about 5 minutes, or until the carrot and potato are tender.

Remove about 1/2 cup of the hot soup and mix it with the miso. Add the sprouts to the pot of soup and as soon as it resumes boiling, stir in the miso mixture. Add the tofu, if desired.

Remove from heat and serve hot.

Serves 4.

*If you don't have kombu, flavor the soup with 1/2 cup of vegetarian stock (see p. 75). Add the ginger, bring to a boil, and add the vegetables.

Soy-noodle soup

1 tsp. oil
1 medium carrot, diced
2 medium scallions, chopped
1/2 green pepper, diced
1 small turnip, finely diced
2½ cups water
1 cup vegetarian stock (see index)
1 cup cooked soybeans
1/2 recipe soy noodles (see next recipe)
1 tbs. tomato paste (optional)

Heat the oil and gently sauté the vegetables, until the pepper is limp. Add the water and simmer 10 minutes. Add the stock and the soybeans, and return the pot to the boil. Taste for seasoning and add the tomato paste, if you like.

Cut or break the noodles into "soup-size" pieces. Sprinkle the noodles into the boiling soup, gradually, so that the soup continues boiling. Turn heat to simmer and cook about 12 minutes more, or until the noodles are done.

Add more water if you think the soup is too thick. Very good served with a big dollop of yogurt.

Serves 4.

Soy noodles

1/2 cup whole wheat flour
1/2 cup natural soy powder
1 egg
1 tsp. soy sauce
1/4 cup soy milk

Mix the flour and soy powder together. Mix in the egg. Add the soy sauce and soy milk and mix vigorously. Turn out onto a floured board and knead for about 5 minutes until dough is smooth and elastic.

Pat the dough into a flat cake and place on a well-floured board. Roll with a well-floured rolling pin, as thinly as possible. (Flour the dough well and wrap it around the rolling pin. Roll. Unwrap the dough, flour it, wrap it around again, and continue in this manner until the dough is as thin as you wish.)

Fold the sheet of dough back and forth upon itself accordion style. Shred with a very sharp knife. The noodle can be any width you like.

To cook: sprinkle the noodles into boiling, salted water to cover. Cook 5 to 8 minutes, or until done to taste.

Uncooked noodles may be refrigerated in an airtight container for several days.

Makes about 3 cups of cooked noodles. (Recipe may be doubled.)

Great soybean soup

6 *cups cooked soybeans*
3 *tbs. vegetable oil*
3 *onions, chopped*
2 *carrots, chopped*
1 *green pepper, chopped*
2 *cloves garlic, minced*
3 *large tomatoes, peeled, seeded and chopped*
2 *tbs. molasses*
sea salt and pepper to taste

Heat the vegetable oil in a large kettle. Gently sauté the onions, carrots, pepper and garlic until the onion is becoming translucent. Add the tomatoes, molasses and the cooked beans. Stir in enough water to make a thick soup.

Cook 30 minutes to blend all flavors. Add salt and pepper to taste and cook 10 minutes more. Serve hot.

This soup may be as thick or thin as you please; add water to thin.

Serves 6, generously.

Soy-tomato soup

2 tbs. oil

1 onion, cut up

1 ½ cup tomatoes, peeled, seeded and chopped

1/2 tsp. rosemary

2 cups cooked soybeans

2 tbs. nutritional yeast

2 cups water

1/4 cup soy sauce

Heat oil and sauté onion until translucent. Add tomatoes and simmer 2-3 minutes. Add all remaining ingredients and bring to a boil.

Serves 4.

Cream of asparagus soup

1½ lbs. fresh asparagus
2 tbs. vegetable oil
1 tbs. minced shallot or onion
4 tbs. soy powder
1 tbs. liquid lecithin

Trim and discard tough part of asparagus stems. Cut asparagus stalks into 1-inch sections. Drop in boiling, salted water to cover and simmer until the color brightens. Remove the asparagus as soon as it is tender. Reserve the cooking water. Separate the asparagus tips from the stalks and set the tips aside.

Add enough water to the cooking liquid to make 3 cups. Puree the cooked asparagus stalks in a blender, using 2 cups of the cooking liquid. Set puree aside.

Heat the oil and sauté the shallot or onion. Stir in the soy powder, mixing thoroughly. Stir in the remaining cup of cooking liquid gradually, stirring with a wire whisk. Bring to a boil and simmer 5 minutes, then whisk in the liquid lecithin. Stir in the puree. Add salt and pepper to taste. Add the reserved asparagus tips and heat through. Stir very gently.

Serves 4.

Ineffable soup

1 *cup Ten-minute tomato sauce* *(see next recipe)*

3 *cups soy milk*

pinch dry, powdered orange peel

sea salt to taste

Puree the tomato sauce in a blender. Bring soy milk to a simmer, add orange peel and tomato puree. Add salt to taste.

Serves 4.

Ten-minute tomato sauce

2 tbs. olive oil

1 medium onion, chopped

1 clove garlic, minced

1 lb. tomatoes (preferably Italian) peeled, seeded, and chopped

1/2 tsp. basil

pinch powdered fennel

Heat the oil in a large skillet. Add the onion and garlic and sauté until the onion is translucent. Add the remaining ingredients and cook at simmer for 10 minutes, stirring occasionally.

Makes about 1½ cups.

Vegetarian stock pot

Stock or bouillon is an important flavoring ingredient in many dishes. Prepared bouillons, cubes and powders are convenient, but are usually extremely salty, as well as containing such additives as monosodium glutamate. We find the following recipe fairly quick and easy to make, dark in color and rich in flavor. We often freeze it in an ice-cube tray, then store the frozen cubes in a freezer container. One or more frozen cubes may be tossed into a dish for extra flavor. Dilute to taste for a simple broth.

2 tbs. oil
1/2 medium onion, chopped (leave the yellow skin on)
1/2 stalk celery, chopped
1 small carrot, chopped
1 medium turnip, chopped (don't peel)
1 clove garlic, chopped
1 small tomato, chopped
few celery leaves, chopped
3-4 sprigs parsley, chopped
1/4 cup nutritional yeast
1/4 cup soy sauce
3 cups water
1/4 tsp. thyme
1/4 tsp. basil

Heat the oil and sauté the onion, celery, carrot and turnip until the onion becomes translucent. Add the garlic, tomato, celery and parsley and cook, stirring, until the mixture resembles a thick paste. Mix in the yeast and soy sauce and stir until smooth. Add the remaining ingredients and simmer over very low heat 15 minutes. Let cool.

Strain the broth, and discard the vegetables. Stir well before using.

Makes about 3 cups of stock.

Cream of celery protein soup

leaves of 1 head of celery
1 *carrot, sliced*
1 *onion, in chunks*
1 *clove garlic*
1 *potato, in chunks*
4 *cups water*
2 *tbs. tahini*
1 *cup soy protein isolate*
1 *cup chopped celery stalks*
3 *tsp. salt*
dash pepper

Put celery leaves, carrot, onion, garlic and potato in large pot. Add water. Bring to a boil and simmer 35 minutes. Drain. Reserve the broth and the potato. In a blender, combine the potato, 2 cups of reserved broth, the tahini, and the soy protein isolate. Blend until smooth. Set aside. Add the chopped celery, salt and pepper to the remaining broth. Bring to a boil and simmer 10 minutes or until celery is just tender. Add mixture from blender. Heat gently, stirring to mix well.

Serves 4.

Soy breakfast sausage

1/2 cup cornmeal
1/2 cup soy granules
1/2 cup buckwheat flour
1½ tsp. sage
1½ tsp. black pepper
1/2 tsp. powdered garlic
3 cups water

Mix dry ingredients with 1 cup of the water. Heat the remaining 2 cups of water to a boil, and gradually stir in the corn-soy mixture. Simmer over low heat, stirring, for about 10 minutes. Pour mixture into oiled loaf pan and chill thoroughly.

To cook, slip the mixture from the mold and cut in 1/4-inch slices. Sauté in a little oil over medium heat until brown and crisp on both sides.

Instant breakfast drink

1 *cup 2%-fat milk*
3 *tbs. non-fat dry milk*
1 *tbs. nutritional yeast*
1 *egg*
1 *tbs. honey*
2 *tbs. carob powder*
2 *tbs. soy protein isolate*

Mix all ingredients in a blender.

Provides about 37½ grams of protein and 522 calories.

Soy scramble

1/2 small tomato, cut up
1/4 small onion, cut up
1 tbs. soy powder
1 tbs. plain yogurt
2 tsp. wheat germ
2 tsp. nutritional yeast
2 eggs
1/2 tsp. chili powder
pinch sea salt
oil

Mix all ingredients except oil in a blender until well mixed.

Heat oil in a large skillet. Add the egg mixture and stir gently until done to taste.

Serves 1.

Tofu and egg

2/3 cup mashed tofu
1 tsp. whole wheat flour
3 dried Chinese mushrooms, soaked, drained and shredded
1 tsp. soy sauce
1 tbs. fresh or frozen green peas
2-3 eggs, lightly beaten
1 tbs. oil
soy sauce.

Mix the tofu with the flour, mushrooms, 1 tsp. soy sauce, peas and eggs. Heat the oil in a large skillet. Add the tofu mixture. Cover the skillet and cook over medium heat for about 5 minutes or until mixture is set. Cut into squares. Serve with soy sauce.

Serves 2.

Super pancake mix

3 cups whole wheat flour

1 cup soy flour

1 cup buckwheat flour

1 cup wheat germ

1 cup roasted peanut flour (optional)

1/2 cup cornmeal

1/2 cup brown rice flour

1/2 cup nutritional yeast

1/2 cup powdered lecithin

Mix all of the above ingredients together. Keep in the refrigerator or freezer until used.

We made the peanut flour optional because it might be hard to find, but it is definitely worth hunting for.

Super pancakes

1 cup super mix
pinch sea salt
1½ tsp. baking powder
2 tbs. oil
2 eggs
3/4 cups skim milk (about)

Mix all of the ingredients together.

Cook on a hot oiled griddle.

Serves 4.

SUPER WAFFLES

Same as super pancakes, but increase the milk to 1 cup (about) and cook on a waffle iron.

Cooler

1 *carrot, cut up*
2 *tbs. chopped parsley*
3 *tbs. soy powder*
1 *tbs. nutritional yeast*
1 *tbs. bran powder**
1 *tbs. lemon juice*
pinch cinnamon
2 *cups tomato juice*
1 *cup sparkling mineral water*

Reserve the mineral water. Place all other ingredients in a blender and blend until smooth. Pour into a serving pitcher and stir in the mineral water. Serve chilled.

Makes about 3 cups.

*Grind bran flakes in a blender to make a powder.

Seven

MAIN DISHES

Soybeans Chinoise

2 tbs. oil
4 slices fresh ginger root
1 stalk celery, thinly sliced
1 carrot, thinly sliced
1 onion, quartered and thinly sliced
2 cloves garlic, crushed
4 tbs. soy sauce
2 cups cooked soybeans
1/2 cup sherry or stock

Heat the oil in a large skillet or wok. Add the ginger, and stir for a minute. Add the celery, carrot and onion, and stir until the vegetables are just beginning to wilt. Add the garlic and stir a minute more. Stir in the remaining ingredients.

Cover the pot and simmer over low heat for 10 minutes. Remove the cover and raise the heat for a minute to reduce the sauce. Stir.

Remove the ginger. Serve hot.

Serves 4.

Chinese stir-fried soy sprouts with vegetables

2 tbs. oil
1 clove garlic, split
1/2 tsp. sea salt
2 stalks bok choy, sliced (or 3 stalks celery)
1 small onion, sliced
1 tbs. chopped green pepper
1/2 tsp. grated fresh ginger
1 cup soybean sprouts
2 tsp. arrowroot or cornstarch
1/2 cup water
1 tbs. soy sauce
1/2 cup tofu, diced
1/2 tsp. sesame oil (optional)

Heat the oil in a large skillet or wok. Add the garlic. Let the garlic brown, then discard it.

Add the salt. Toss in the bok choy, onion, pepper and ginger. Stir over high heat until the onion is becoming translucent. Stir in the sprouts. Stir for about 1 minute.

Mix the arrowroot with the water, add the mixture to the vegetables and bring rapidly to a boil, stirring. Stir in the soy sauce. Add the tofu and stir very gently, just to heat.

Remove to a serving dish and garnish with the sesame oil.

Serves 2.

Tofu with vegetables

1 tsp. vegetable oil
1/4 lb. mushrooms, sliced
1 carrot, sliced
1 small onion, sliced
1 cup fresh mung bean sprouts
2 cloves garlic, crushed
1/2 tsp. sea salt
1 cake tofu, cut in cubes
3 tbs. soy sauce
2 tbs. sesame tahini
1/2 tsp. honey
1 tbs. tomato paste
1/4 tsp. freshly ground ginger

Heat the oil in a wok or heavy skillet. Add the mushrooms, carrots, and onion. Stir vigorously for about 2 minutes.

Add all of the remaining ingredients. Cover the pot and steam for 3 minutes. Serve hot.

Serves 2.

Tofu scramble

2 tbs. oil
1 medium onion, diced
1 cup diced fresh mushrooms
1 cup diced tofu
2 tbs. fresh or frozen peas
2 tsp. honey
2 tbs. soy sauce

Heat the oil. Sauté the onion and mushrooms until the onion is translucent. Add the tofu and peas and stir to mix well. Mix together the honey and soy sauce. Add to the tofu mixture. Stir to heat thoroughly. Taste for seasoning and add more soy sauce if desired. Serve with rice.

Serves 2.

Tofu curry

2 *tbs. oil*
2 *cakes tofu, diced*
1 *large onion, halved and sliced*
4 *tomatoes, peeled, seeded and chopped*
1 *tsp. grated ginger*
1 *clove garlic, minced*
1 *tbs. curry powder*
1 *cup water*
1/2 *tsp. salt*
1/2 *tsp. honey*
2 *tsp. cornstarch*
1 *tbs. soy sauce*

Heat 1 tbs. of the oil in a large skillet and sauté the tofu gently until lightly golden. Remove from the skillet and set aside. Add the remaining oil to the skillet and sauté the onion for a minute, then add the tomatoes, ginger, garlic, curry powder, water, salt and honey. Cook, stirring, for about 5 minutes, or until the onion is soft. Mix the cornstarch and soy sauce together, and add to the curry. Stir. If the mixture is too thick, add a little water. Add the tofu to the curry mixture and stir gently, just to heat. Serve over rice or noodles.

Serves 2.

Eggs fu yong

1 *cup steamed soy sprouts*
liquid from sprouts plus enough water to make 3/4 cup

1 *tsp. tomato paste*
2 *tbs. soy sauce*
1/2 *tsp. honey*
1 *tbs. cornstarch mixed with 2 tbs. water*
3 *scallions, minced (separate white and green parts)*
6 *eggs*
pinch sea salt
vegetable oil (sesame oil is good)

Bring the 3/4 cup of liquid to a boil. Add the tomato paste, soy sauce and honey. Add the cornstarch mixture to the boiling liquid while stirring. Add the green part of the scallions. Set aside.

Beat the eggs. Add the white of the scallions, salt and the sprouts. Mix well.

Heat a little oil in a skillet or wok. Add about 1/6 of the egg mixture (about 1/3 cup). Cook on one side; turn the patty and cook on the other side. Remove to a warm platter. Continue until all of the egg mixture is cooked. Top with the warm sauce.

Serves 6.

Beans and pasta Neapolitan

2 tbs. olive oil
2 tbs. vegetable oil
2 cloves garlic, split
4 cups canned tomatoes
1 tsp. oregano
2 tsp. chopped parsley
1 ½ cups cooked soybeans
1/2 lb. whole wheat pasta
(preferably "elbows")
2 tsp. chopped fresh basil (1/2 tsp. dried)

Heat the 2 oils in a small skillet. Add the garlic and cook until the garlic is browned. Remove and discard the garlic. Set the oil aside.

Put the tomatoes in a large, heavy pot. Add the oregano, parsley, and garlic-flavored oil. Cook over moderate heat, stirring occasionally, for about 40 minutes, or until the sauce is quite thick. Add the beans and simmer 10 minutes more.

Meanwhile, drop the pasta into boiling, salted water and cook until it is just "al dente," i.e., it still offers some resistance to the teeth. Drain and set aside.

Mix the cooked pasta into the completed bean mixture, and heat thoroughly. Serve hot, garnished with the basil.

Serves 4.

Chili loaf

3 cups cooked soybeans
1 small green pepper, minced
1 small onion, minced
1 tbs. oil
1½ cups fine, dry, whole wheat bread crumbs
2 cloves garlic, crushed
1½ cups grated sharp cheddar cheese
1 tbs. chili powder (or more, to taste)
1 tsp. sea salt
1/4 tsp. basil
1/4 tsp. oregano
1/2 tsp. cumin powder
1/4 tsp. celery seed

Heat oven to 350° F. Oil a loaf pan or casserole.

Mix all ingredients thoroughly and press the mixture lightly into the prepared pan.

Bake at 350° F. for one hour or until browned nicely. Serve hot or cold.

Chili sin carne

1 cup dried soybeans

1 cup lentils

1/2 cup sesame seeds

2 green peppers, chopped fine

3 medium onions, chopped fine

2 tbs. peanut oil

1 tbs. salt

2 tbs. chili powder

2 cloves garlic, crushed

1/2 tsp. powdered cumin

1/2 tsp. oregano

Cover the soybeans with cold water. Bring to a boil, remove from heat and let stand one hour. Drain the beans, cover with fresh cold water, bring to a boil and simmer over low heat 1 hour.

Add the lentils and sesame seeds to the soybeans. Add enough cold water to cover. Bring to a simmer.

Sauté the onions and peppers briefly in the peanut oil until they are barely limp. Add the vegetables and oil to the bean pot. Add the remaining ingredients, and simmer for about 30 minutes or until everything is tender.

Serve hot with brown rice

Serves 6.

Garnishes for chili dishes

Guacamole, grated cheese, diced tomato, minced onion, shredded lettuce, minced radish.

Vegetable curry

1/4 cup water
1 small zucchini, diced
1 small turnip, diced
1 medium onion, diced
1 stalk celery, diced
2 cups cooked soybeans
1/2 cup black raisins
1 cup curry sauce (see p. 97)
1/4 cup sesame seeds
1/4 cup freshly grated coconut

Bring the water to a boil and add the diced vegetables. Cover the pot and steam the vegetables over low heat for 10 minutes. Add the soybeans raisins and curry sauce. Bring to a boil and simmer 5 minutes.

Meanwhile, place the sesame seeds and coconut in an uncoated iron skillet. Toast over moderate heat, stirring, until the sesame seeds are beginning to brown. Remove the seeds and coconut from the pan immediately. Set aside.

Place the curry mixture in a serving dish and top with the sesame coconut mixture.

Serve with brown rice. Add lemon wedges to season, if desired.

Serves 4.

Curry sauce

2 tbs. oil
4 tbs. chopped onion
1 clove garlic, crushed
3 tbs. whole wheat flour
3 tsp. curry powder
1⅓ cups apple juice
1/2 cup yogurt
1 tsp. sea salt
3 tbs. nutritional yeast

Sauté the onion and garlic in the oil until the onions are translucent. Stir in the flour and stir for a minute or two. Stir in the curry powder.

Add the apple juice gradually, stirring to make a smooth sauce. Stir in the yogurt, salt and yeast.

Simmer for about 5 minutes, or until the sauce is thickened.

Use in recipes as indicated, or to stretch leftover fruit or vegetables.

Makes about 1½ cups.

Casserole Printanier

3 cups cooked soybeans
2 stalks celery, diced
2 carrots, diced
1 medium onion, diced
1 turnip, diced
1 green pepper, diced
2 tsp. basil
3 tbs. minced parsley
1/3 cup whole wheat flour
1 tbs. sea salt
dash pepper
1 cup hot stock or water
1 tbs. molasses
2 tbs. lemon juice
wheat germ

Heat oven to 350° F. Oil a 1½-qt. casserole.

Mix the beans with the celery, carrots, onion, turnip, green pepper, basil and parsley. Mix the flour with the salt and pepper and toss the flour mixture with the vegetable mixture.

Put the vegetable-flour mixture into the casserole. Mix the stock with the molasses and lemon juice, and pour over the vegetable mixture.

Bake covered for 30 minutes. Uncover. Sprinkle liberally with wheat germ and bake another 15 to 20 minutes, or until the top is brown and bubbly.

Serves 4.

Vegetable and cheese casserole

1/2 cup chopped parsley
1/3 cup whole wheat flour
1½ tsp. sea salt
1/4 tsp. basil
1/4 tsp. thyme
4 oz. cheddar cheese, grated
1/2 cup soy granules
2 carrots, sliced
1 onion, sliced
2 turnips, sliced (about 1 lb.)
2 small potatoes, sliced (about ½ lb.)
1 cup fresh or frozen peas
1 green pepper, chopped fine
1/4 lb. mushrooms, sliced
1⅔ cups stock or water
soy granules.

Heat oven to 350° F. Oil a large (4 qt.) casserole.

Mix together the parsley, flour, salt, herbs, cheese and 1/2 cup of soy granules.

Arrange a layer of vegetables in the casserole, using about 1/3 of the vegetables. Sprinkle with 1/3 of the parsley mixture. Repeat these layers twice more. Pour the stock over all. Sprinkle the top with soy granules.

Cover the casserole with a lid or foil and bake 35 minutes. Remove the cover and bake 15 minutes more.

Serves 4 to 6.

Granule and vegetable casserole

1/2 cup soy granules
3/4 cup water
1/4 cup soy sauce
1 onion, minced
3 tbs. sesame seeds
2 tbs. nutritional yeast
1/2 tsp. oregano
1 tsp. basil
1/2 cup grated mozzarella cheese
2 medium zucchini, sliced
2 medium tomatoes, sliced
3/4 cup tomato juice

Oil a large casserole dish. Heat oven to 350° F.

Cook the soy granules with the water and soy sauce until nearly all of the water is absorbed. Toss with the onion, sesame seeds, yeast, oregano, basil and cheese. Place a layer of zucchini in the bottom of the casserole and add a layer of tomatoes. Top with half of the granule mixture. Add the remaining zucchini and tomatoes and top with the remaining granule mixture. Pour the tomato juice over all. Bake 1½ hours.

Serves 4 to 6.

Savory stuffed cabbage

1 cup soy granules	*1/8 tsp. black pepper*
1½ cups stock or water	*1 tbs. soy sauce*
2 tbs. oil	*1/2 cup minced ripe olives*
1 medium onion, minced	*3/4 cup grated cheddar cheese*
1 tsp. sea salt	*1 small head cabbage*
1/8 tsp. cinnamon	*1 cup tomato juice*

Add the soy granules to the 1½ cups of stock and let soak 4 to 5 minutes. Heat the oil and sauté the onion until it is translucent. Mix together the salt, cinnamon, pepper, soy sauce, olives and cheese.

Drop the cabbage into boiling water to cover until the color brightens. Remove the cabbage from the boiling water, rinse with cold water. Peel away the limp outer leaves. Depending on size, you will probably need about 12 leaves. If there are not enough limp leaves, return the still crisp cabbage to the boiling water and repeat the above procedure.

Mix the soaked granules with the onion, oil and cheese mixture. Place 2 to 4 tbs. of the mixture in the center of a cabbage leaf. Roll up the stem end to cover the filling, fold the sides of the leaf in toward the center, and roll the whole bundle toward the tip of the leaf. Place the stuffed leaves, seam side down, in a large pot. Make two layers, if necessary.

Add the tomato juice, cover the pot tightly, and cook at barely simmering for 45 minutes. Add more liquid, if needed.

Serves 4.

New England baked soybeans

2 cups dried soybeans
1 medium onion
1 tsp. sea salt
1/2 cup dark molasses
1/4 cup honey
1/2 tsp. dry mustard
2 tbs. vegetable oil
1/4 cup soy sauce

Cover beans with water. Bring to a boil and simmer 2 minutes. Cover and let stand 1 hour. Return to the boil and simmer until tender (about 1 hour). Add liquid as needed. Drain the beans, reserving the liquid.

Put the beans in a large, heavy pot. Heat oven to 300° F. or use an electric bean pot.

Cut the onion into chunks and bury them in the beans. Mix the remaining ingredients with the reserved bean liquid and pour it over the beans.

Cover the pot and bake at 300° F. for 6 to 8 hours. Add boiling water as needed.

Serves 6.

Garden burgers

1 carrot
1 small zucchini
1 small onion
3 tbs. soy granules
2 tbs. wheat germ
1/2 tsp. sea salt
2 tbs. cornmeal
3 eggs
1 tbs. minced parsley
vegetable oil (about 2 tbs.)

Grate the carrot, onion and zucchini together. Mix with the soy granules, wheat germ, salt and cornmeal.

Separate two of the eggs. Beat the two separated yolks with the whole egg and add to the vegetable mixture. Beat the two remaining whites until stiff, but not dry. Add the parsley to the vegetable mixture. Gently fold in the egg whites.

Oil and heat a griddle or heavy skillet. Drop vegetable mixture by teaspoonfuls onto the griddle. Cook slowly, about 4 minutes on each side.

Vary the flavor by adding fresh minced herbs in season.

Makes about 18 small patties. Serves 6.

Soybean patties

1 cup dry soybeans
1 tbs. chopped parsley
1 small onion, minced
1/3 cup wheat germ
2/3 cup whole wheat flour
1/4 cup tamari soy sauce
1 tsp. sea salt
1/2 tsp. thyme
1/2 tsp. basil
1/2 tsp. celery seed
1/4 cup cooking water from the beans (or stock or water)
vegetable oil

Soak and cook the beans. Drain them, reserving the liquid.

Mash or chop the beans and mix in all the remaining ingredients except the oil.

Heat a heavy skillet, and cover the bottom with a small amount of vegetable oil. Form patties of the bean mixture and cook them over very low heat until crisp and brown, turning them once. The patties will be rather fragile and crumbly to start, but just handle them gently, they become more cohesive as they cook.

Makes about 8 patties.

Soy loaf

1 stalk celery, grated
1 carrot, grated
1 onion, grated
2 tbs. oil
1/4 cup raw wheat germ
2 tbs. tamari soy sauce
1/2 tsp. sea salt
2 eggs
2 cups cooked soybeans
1 tbs. molasses
1/2 cup tomato juice

Heat oven to 425° F.

Sauté the celery, carrot and onion in the oil for about five minutes or until the vegetables are quite limp. Add the wheat germ and contnue sautéing for about one minute more.

Beat together the tamari, salt and eggs. Mash the soybeans with the molasses and tomato juice. (The beans, molasses and juice may be whirred in a blender if you prefer.) Mix the vegetable mixture, the egg mixture and the bean mixture together.

Place in an oiled loaf pan. Bake at 425° F. for 30 to 40 minutes or until lightly browned.

Serves 4 to 6.

Super-delicious soy-rice patties

1/2 cup soy granules

1 cup cooked brown rice

1 tbs. finely chopped scallion

3 tbs. chopped parsley

1/4 tsp. thyme

1 tbs. nutritional yeast

1/4 cup soy sauce

3/4 cup soy yogurt (or plain dairy yogurt)

2 eggs, lightly beaten

Mix all ingredients together thoroughly. Let stand 10 minutes. Form small patties.

Fry in vegetable oil over moderate heat until golden brown and crisp on both sides.

Serve as is, or top with a well-seasoned soy béchamel sauce (see p. 49).

Serves 4.

Soy hash

2 tbs. oil

1/3 to 1/2 cup chopped onion

1½ cups cooked, drained soybeans

1 tbs. tomato paste

1 tbs. nutritional yeast

1/4 cup soy sauce

2 tbs. tahini

2 cups cooked potatoes

1/3 cup potato cooking water or stock

1/3 cup chopped parsley

sea salt and pepper to taste

Heat oil. Add onion and soybeans. Mash beans lightly. Sauté, stirring frequently, for about 3 minutes.

Add tomato paste, yeast and soy sauce. Mix well. Cook over low heat for about 5 minutes, stirring occasionally.

Add tahini and mix. Add potatoes and mash lightly. Add water and parsley and stir until water is absorbed. Season to taste.

Serves 4.

Unbelievable spaghetti sauce *(People don't believe it's meatless!)*

2 tbs. oil
1 onion, minced
1 carrot, minced
1 stalk celery, minced
1-2 cloves garlic, crushed
1/2 cup chopped pecans
2 tbs. soy powder
1 tbs. nutritional yeast
3/4 cup soy granules
1/3 cup soy sauce
2 cups canned tomatoes, drained and chopped
liquid from the tomatoes plus enough water to make 1 ½ cups
1 tsp. basil
pinch ground fennel
pinch oregano
3 tbs. tomato paste
sea salt and pepper to taste

In a large pot, heat the oil and sauté the onion, carrot, celery and garlic until they wilt.

Add the pecans and stir briefly. Stir in the soy powder and yeast, mixing well.

Add the granules, soy sauce, tomatoes, 1½ cups of liquid, basil, fennel and oregano. Bring to a boil.

Stir in the tomato paste. Simmer 15 minutes. Add salt and pepper to taste.

Makes about 1 quart.

Soy pizza

Crust

2/3 cup water at 110° to 115° F.
1 tbs. dry yeast granules
pinch powdered ginger
1 cup whole wheat flour
1/4 cup soy powder
3 tbs. powdered bran
1 tbs. nutritional yeast
1/2 tsp. sea salt

Add the yeast granules and the ginger to the water. Cover and set in a warm place for about 5 minutes, or until the yeast is becoming frothy. Stir in the remaining ingredients.

Knead on a well-floured board for 5 minutes. Spread the dough on an oiled pizza pan or cooky sheet by pressing the dough outward from the center. Cover and let rest in a warm place for about 10 minutes.

Heat oven to 450° F.

Topping

1 cup Unbelievable spaghetti sauce (see p. 108)
1 medium onion, diced
6-7 mushrooms, sliced
1 tbs. oil
1 tsp. oregano
1/2 tsp. sea salt
1/4 cup freshly grated Parmesan cheese
1/2 cup freshly grated mozzarella cheese

Spread the sauce on the pizza. Heat the oil and sauté the onion and mushrooms until limp. Spread over the sauce. Sprinkle on the remaining ingredients in order.

Bake the pizza on the oven floor for about 10 minutes, or until the cheeses are melted and bubbling.

Lasagna

1 lb. lasagna noodles, preferably whole wheat
3 cups Unbelievable spaghetti sauce (see p. 108)
2 tbs. tomato paste
3/4 cup water
2 eggs, lightly beaten
2 cups ricotta cheese (or drained cottage cheese)
1/2 cup grated Parmesan cheese
1 lb. thinly sliced mozzarella cheese

Cook the noodles in a large amount of boiling, salted water for 5 minutes, or until done. Heat the spaghetti sauce with the tomato paste and the 3/4 cup of water.

Heat oven to 350° F. Oil a large baking dish.

Mix the eggs, ricotta and Parmesan. Place half of the cooked noodles in the baking pan. Spread with half of the ricotta mixture, then half the spaghetti sauce and layer with half the sliced mozzarella. Repeat with the remaining noodles, ricotta, sauce and mozzarella.

Bake 30 minutes, or until bubbly. Let stand 10 minutes before serving.

Serves 6.

Eggplant Parmesan

1 *large eggplant, peeled and sliced*
oil
1/3 *cup soy granules*
2 *cups tomato sauce*
1/3 *cup freshly grated Parmesan cheese*
1/3 *cup freshly grated mozzarella cheese*

Heat oven to 350° F. Oil an 8-inch by 8-inch baking dish.

Sauté the eggplant slices in the oil until both sides are lightly browned. Place a layer of sautéed eggplant in the bottom of the baking dish and sprinkle with half the granules. Top with the remaining eggplant and the remaining granules. Pour the tomato sauce over all, and sprinkle on the cheeses.

Bake 25 to 30 minutes, or until the cheeses are melted and the sauce is bubbly.

Serves 2 as a main dish, or 4 as a side dish.

Stuffed green peppers

6 medium-size green peppers
1/2 cup soy granules
3/4 cup tomato juice
1/4 cup soy sauce
2 cups cooked brown rice
3 tbs. chopped parsley
1 tsp. basil
2 tbs. chopped onion
black pepper

Heat oven to 350° F.

Halve peppers lengthwise, remove stems, pulp and seeds. Dip into lightly salted boiling water for 5 minutes. Rinse in cold water and drain. Mix all remaining ingredients together and fill pepper halves with the mixture. Place, stuffing side up, in a shallow pan. Add 1/2 cup water and bake for 30 minutes or until tops are lightly browned.

Serves 6.

Frijoles Havana

1 tbs. oil
1 tsp. olive oil
1 onion, minced
1 green pepper, minced
1 large clove garlic, minced
1/2 tsp. oregano
2 tbs. vinegar
1/4 cup water
2 cups cooked soybeans
sea salt and pepper to taste

Heat the oils and sauté the onion, pepper and garlic until quite soft. Add the oregano, vinegar, water and soybeans.

Mash the soybeans slightly, cooking over moderate heat until all the liquid is absorbed.

Serve with brown rice.

Makes 2 cups.

Mexi-bean pie

Crust

1/4 cup soy powder
3/4 cup cornmeal
1 cup water
1/2 tsp. basil
dash cayenne
2½ cups boiling water
1/2 tsp. sea salt

Mix together the soy powder, cornmeal, 1 cup water, basil and cayenne.

Add the salt to the 2½ cups of boiling water and stir in the cornmeal mixture. Partially cover the pan and cook for 20 minutes, stirring occasionally. Uncover and cook 15 minutes more, stirring often. Let cool a few minutes.

Heat oven to 425° F. Oil a deep-sided 9-inch pie pan.

Spread the cornmeal mixture on the bottom and sides of the pan. Bake 15 minutes. Remove the dish from the oven and allow to cool while you prepare the filling.

Filling

2 tbs. oil
1 medium onion, minced
2 cloves garlic, minced
1 green chili (fresh or canned) chopped
1½ cups cooked soybeans
2 tbs. tomato paste
2 tsp. chili powder
1 tsp. cumin
1/2 tsp. oregano
1 tsp. sea salt
2 tbs. tahini
3/4 cup grated cheddar
1 egg, beaten

Mexi-bean pie 2

Heat the oil in a large skillet. Sauté the onions and garlic until the onion is translucent.

Stir in the green chili, soybeans, tomato paste, chili powder, cumin, oregano and salt. Stir to mix well, and heat gently. Stir in the tahini.

Remove from heat and stir in the cheddar. Stir in the egg.

Heat oven to 350° F. If the sides of the crust have slipped down the sides of the pan, spread the cornmeal batter upward. Pour the filling into the crust. Spread smooth.

Bake at 350° F. for 20 to 25 minutes. Let stand 15 minutes before cutting.

Eight

BREADS ETC.

Soy biscuits

3/4 cup whole wheat flour
1/4 cup soy powder
1/4 cup wheat germ
1 tsp. baking powder
2 tbs. oil
pinch sea salt
1/3 cup yogurt (about)

Heat oven to 450° F. Oil an 8-inch round or square baking pan.

Mix together the flour, soy powder, wheat germ and baking powder. Mix well. Stir in the oil and salt. Add the yogurt. The dough should be a little stiff.

Drop tablespoonfuls of the dough into the prepared pan. Shape a little with your fingers, if you like.

Bake for 10 to 12 minutes, or until lightly browned.

Herb breadsticks

1¼ cups soy milk, scalded and cooled to 115° F.

2 tbs. honey

1 tbs. dry yeast granules

1 tbs. vegetable oil

1½ tsp. sea salt

2 tsp. dried basil

1/2 tsp. powdered fennel

1 tsp. dried oregano

1/2 tsp. pepper

1/2 cup soy powder

2½ cups whole wheat flour (about)

Add the honey and yeast to the milk and let rest about 10 minutes or until the yeast is bubbly.

Add the oil, salt, herbs and pepper to the yeast mixture. Stir to mix. Add the soy powder and 2 cups of flour and mix well. Add the remaining flour, a little at a time, until the dough can be handled. Knead for 10 minutes on a well-floured board. Place the dough in an oiled bowl, turning to coat the dough with oil. Cover with a cloth and set to rise for 45 minutes or until doubled in bulk.

Punch the dough down. Let it rest on a floured board for 10 minutes. Knead 3 to 4 times.

Divide the dough in half. Roll each half into a rectangle about 12 inches wide. Cut each rectangle into 12 sections. Roll each section between floured hands to make a baton about 1/3-inch around and 12 inches long.

Place the batons on an oiled cooky sheet. Cover with a cloth and allow to rise until doubled in bulk, about 45 minutes. During the last 10 minutes of rising, preheat the oven to 400° F.

Bake the breadsticks for about 15 to 20 minutes or until done.

Soy wafers

3/4 cup soy powder
1 tsp. seasoned salt
1 egg, beaten
1 tbs. soy milk
1/3 cup grated sharp cheese
1/4 cup ground or finely chopped brazil nuts (or other nuts or seeds)

Heat oven to 350° F. Oil a baking sheet.

Mix soy powder and salt. Stir in egg and soy milk. Stir in cheese, blending thoroughly.

Flour a board with soy powder, and roll the dough to about 1/4-inch thickness. Sprinkle with the nuts and roll lightly to press in the nuts. Cut into 1½-inch squares.

Place on oiled sheet and bake for about 20 minutes, or until lightly browned.

Makes about 2 dozen.

Spoon bread

3/4 cup yellow cornmeal
1/4 cup soy granules
1/2 tsp. salt
2 cups water
2 eggs, separated
1 cup milk
2 tbs. oil

Oil an 8-inch by 8-inch baking dish. Heat oven to 350° F.

Mix the cornmeal, granules, salt and water. Bring to a boil and simmer 5 minutes, stirring constantly. Mix the egg yolks with the milk and oil and add to the cornmeal mixture. Beat the egg whites until stiff but not dry and fold them into the mixture. Pour into the prepared baking dish and bake 1 hour. Serve hot from the dish. (Use a spoon, of course.)

Soya-rye muffins

1 cup soya powder
1 cup rye flour
1/4 cup bran
1 tbs. baking powder
1/2 tsp. sea salt
1/4 cup honey
1 egg, beaten
2 tbs. oil
1 cup water

Heat oven to 350° F.

Mix dry ingredients together. Mix together the honey, egg, oil and water. Add the liquid mixture to the dry ingredients, stirring to mix well.

Fill oiled muffin tins to about 3/4 full. Bake for 35 to 40 minutes.

Makes 12 muffins.

Corn meal-soy muffins

1/4 cup soy powder
1/2 cup whole wheat flour
1¼ cups yellow cormeal
3 tsp. baking powder
3/4 tsp. salt
1 cup milk
2 eggs, lightly beaten
4 tbs. oil

Heat oven to 400° F. Oil a 12-muffin tin.

Mix dry ingredients together. Mix milk, eggs and oil and add to the dry ingredients, mixing well. Fill muffin tins about 2/3 full. Bake for 25-30 minutes.

Makes 12 muffins.

Quick soy bread

2½ cups whole wheat flour
2/3 cup soy powder
2 tsp. baking soda
1/2 tsp. salt
2 cups soy milk
1 tbs. lemon juice
2 tbs. oil
1/4 cup honey

Oil a 9-inch by 5-inch by 3-inch loaf pan. Heat oven to 325° F.

Mix together all the dry ingredients. Mix the liquid ingredients together and add them to the dry ingredients, mixing well. Smooth the batter into the loaf pan and bake for 1 hour or until a cake tester comes out clean.

Whole wheat bread

3 cups skim milk, scalded and cooled to 115° F.
1/2 cup molasses
pinch powdered ginger
6 tbs. dry yeast granules
1/4 cup vegetable oil
1 cup soy powder
1/2 cup non-instant dry milk powder
1/2 cup lecithin powder
1/2 cup nutritional yeast powder
1 cup gluten flour
1/2 cup wheat germ
2 tsp. sea salt
8 to 12 cups whole wheat flour, approximately

NOTE: All ingredients should be at room temperature unless otherwise noted.

Stir the molasses and ginger into the milk. Add the yeast and stir just to mix. Cover the mixture with a cloth and let it rest in a warm place for about 10 minutes. At the end of 10 minutes, the yeast should be bubbly. (If it is not, the yeast is dead, and you must begin again.)

Stir the vegetable oil into the yeast mixture.

In a large bowl, mix together the soy powder, milk powder, lecithin, nutritional yeast, gluten flour, wheat germ, sea salt and 4 cups of whole wheat flour. Mix well, then stir in the yeast mixture. Let it rest 10 minutes, as whole wheat flour may absorb moisture slowly, and adding all the flour quickly might result in heavy bread.

Add more flour to the mixture, one cup at a time, until the dough forms a fairly cohesive mass that can be lifted from the bowl.

Flour your kneading board well and place the mass of dough on the flour. Pat it into a roundish shape. Flour your hands. Press the heels of both hands into the dough mass at the center, pushing away from your body. Lift your hands. With both hands, lift the edge of the dough farthest from you, and fold it toward you. It should fold nicely on the crease you made with the heels of your hands. You should now have a rough half-circle of dough with its diameter horizontal to you. Lift the mass of dough and give it a quarter turn so that the diameter is vertical to you. If the dough sticks to the board as you turn, sprinkle flour beneath it. Again, press the heels of both hands into the center of the dough, pushing away. Fold the dough; give it a quarter turn. Flour if you need to. Press, push, fold, turn—that is all there is to kneading, and you will soon develop a rhythm of your own. Add flour whenever the dough gets sticky; you may need to add quite a bit. As you knead, you will feel the dough becoming smooth and springy. It will reach a point where it no longer needs flour at every turn; it is fairly smooth, and quite springy to the touch. At that point you may stop kneading. Roughly, it takes about 10 minutes of kneading to reach this point. Until your experience tells you the dough is done, you may let a clock be your guide.

Oil a large bowl. Pat the dough into a ball and place it in the bowl, turning it to cover all the surfaces lightly with oil. Cover with a cloth and set to rise in a warm place (see above), for about 45 minutes until it has doubled in bulk. To determine whether the dough has doubled, jab two fingers sharply into the center of the dough, pressing down about 1/2 inch. If the impressions remain, consider the dough doubled.

Push a fist into the center of the dough ball—this is "punching down." Bring the edges of the dough toward the center. Flour the kneading board. Turn the dough onto the floured board and let it rest 10 minutes.

While the dough rests, grease 3 breadpans with lard or vegetable oil.

Knead the dough 3 or 4 times to work out the largest air bubbles. With a sharp knife, divide the dough into roughly equal thirds. Push, pat and fold each third until it is somewhat loaf shaped. There will probably be a seam on one side of the loaf, so make sure that that side is at the bottom and place your dough in the bread pan seam-side down.

Cover the three loaves with a cloth, and set them in your rising place for about 45 minutes or until doubled. You can *see* that the loaves have doubled; don't jab them with your fingers, or you will have dented bread.

When the loaves have been rising for 35 minutes, heat the oven to 350° F. (If your loaves were rising in the oven, you'll have to move them.)

Place the loaves in the oven, allowing space for air to circulate around them. Bake for about 45 minutes, or until the top is brown and the loaf makes a hollow sound when you thump it on the bottom.

Remove the bread from the pans immediately and cool on wire racks. Let the bread cool for at least 20 minutes before cutting it.

Makes 3 loaves.

Honey date bread

1 ½ cups whole wheat flour
1/2 cup soy powder
3 tsp. baking powder
1 tsp. salt
1 cup soy granules
2 eggs, lightly beaten
1/4 cup oil
3/4 cup milk
3/4 cup honey
1 cup chopped, pitted dates

Oil a 9-inch by 5-inch by 3-inch loaf pan. Heat oven to 350° F.

Mix or sift together the flour, soy powder, baking powder and salt. Stir in the granules. Mix together the eggs, oil, milk and honey. Add to the dry ingredients and mix well. Add the dates and stir to mix. Smooth the thick batter into the prepared pan and bake for an hour or until the bread tests done in the center.

Bran bread

1 tbs. dry yeast granules
1¾ cups water at 110° F.
2 tbs. honey
1 tsp. sea salt
1 tbs. vegetable oil
1 cup bran flakes
1 cup soy powder
3 to 3½ cups whole wheat flour

Grease a 9-inch bread pan.

Sprinkle the yeast over the warm water and stir in the honey. Cover and let stand in a warm place for about 5 minutes or until the yeast bubbles. (If the yeast doesn't bubble, it is dead and you must start again.)

All other ingredients should be at room temperature. Mix together the salt, oil, bran flakes and soy powder and add to the yeast mixture. Slowly stir in the flour until the dough forms a mass.

Turn the dough out onto a well-floured board and knead for about 5 minutes until smooth and elastic, adding flour as needed to keep the dough from sticking. (See Whole Wheat Bread for detailed kneading instructions.)

Shape the dough into a fat cylinder and put it into the greased pan, seam-side down. Cover with a clean cloth and let rise in a warm place until about doubled in bulk (about 45 minutes).

Heat oven to 350° F.

Gently brush the top of the loaf with oil. Bake the bread for 45 to 50 minutes or until done. Cool on a wire rack.

Makes one large loaf.

Nine

SWEET THINGS

Super squares

7 tbs. vegetable oil
4 tbs. honey
1 tbs. liquid lecithin
1/2 cup dark molasses
2 eggs, lightly beaten
1½ cups whole wheat flour
1/2 cup soy powder
1/3 cup non-instant dry milk
1½ tsp. baking powder
1 tsp. ground cinnamon
1/4 tsp. ground cloves
1/4 tsp. powdered ginger
1 cup raisins
1/2 cup chopped nuts, or seeds or soy granules

Heat oven to 350° F. Oil and flour an 8-inch by 13-inch baking pan.

Mix together the oil, honey, lecithin, molasses and eggs. Mix the dry ingredients together. Add the oil mixture to the dry ingredients, mixing well. Stir in the raisins and nuts.

Place the mixture in the baking pan and bake for 10 to 15 minutes, or until done. Cut in 1-inch squares. Let cool in the pan.

Soy-apple betty

oil
1/3 cup soy granules
1/3 cup wheat germ
1/4 cup honey
1/2 cup warm water
1 tbs. lemon juice
4 medium apples, cored and sliced thin
cinnamon

Heat oven to 350° F. Oil an 8-inch by 8-inch baking dish.

Mix together the granules, wheat germ, honey, water and lemon juice.

Place half of the sliced apples in a layer on the bottom of the baking dish, top with half of the granule mixture. Sprinkle with cinnamon to taste. Add a layer of the remaining apples, top with the remaining granule mixture, sprinkle with cinnamon and cover the dish tightly.

Bake covered for 30 minutes. Remove the cover and bake 5 to 10 minutes more, or until the topping is crisp.

Serves 4 to 6.

Custard with protein plus

1/2 cup soy powder
4 level tbs. soy protein isolate
2 cups water
3 tbs. honey
3 eggs, lightly beaten
pinch sea salt
1/2 tsp. vanilla
1/4 tsp. allspice

Heat oven to 325° F. Oil a 2-quart casserole.

Mix all ingredients together and stir with a wire whisk until smooth. Pour into prepared casserole.

Set casserole in a pan of hot water and bake 2½ hours, or until an inserted knife comes out clean.

Serves 6.

Protein halvah

1 cup sesame seeds

1/4 cup honey

pinch sea salt

6 tbs. soy protein isolate

1 tsp. lemon juice (or more to taste)

Heat the sesame seeds in a heavy iron skillet until they begin to brown lightly. Stir constantly. Remove immediately from the hot pan.

Puree the sesame seeds in a blender to make a rough paste. Remove the sesame paste to a bowl and stir or knead in the remaining ingredients.

Pat the mixture about 1/3-inch thick on waxed paper. Chill. Cut into whatever shape and size you fancy.

Soy cheesecake

Crust

1/4 cup low-fat soy powder
3/4 cup whole wheat flour
1/4 cup honey
1 tbs. oil
1 egg yolk
2 tbs. cold water

Blend all ingredients together. Pat to cover bottom and side of a 9½-inch pie pan. (Dampen hands slightly to spread dough.)

Bake at 350° F. for 5 minutes. Set aside.

Filling

2 whole eggs + 1 egg yolk, beaten lightly
2 cups drained cottage cheese
1/4 cup low-fat soy powder
1/2 cup honey
2 tsp. lemon juice
1 tsp. vanilla
2 egg whites, beaten stiffly

Heat oven to 350° F.

Combine the whole eggs, yolk, cottage cheese, soy powder, honey, lemon juice and vanilla. Placing about half of the mixture in the blender, blend until smooth. Repeat with the remaining half. Fold in the egg whites.

Smooth the filling into the prepared crust and bake for about 35 minutes, or until lightly browned.

Serve warm or cold. Top with fresh berries, if desired.

Fruit "candy"

3 tbs. sesame seeds
1 cup seedless raisins
*1 cup dried apricots**
1 tbs. wheat germ
4 tbs. sunflower seeds
2 tbs. soy granules
2 tbs. grated fresh coconut

Toast the sesame seeds in a heavy, uncoated pan until they "jump." Remove the seeds from the pan.

Chop the raisins and apricots together. Add the sesame seeds, wheat germ, sunflower seeds and soy granules. Add 1 tbs. of the grated coconut. Mix as thoroughly as possible, using a fork.

Form 1-inch balls. Sprinkle with the remaining coconut.

These will keep in the refrigerator for about 1 week. Store in an airtight container.

*Natural, sun-dried apricots work best, as very moist apricots will make a very sticky mixture.

Honey spice cake

1 ½ cups whole wheat flour
1/2 cup rice flour
1/2 cup soy powder
2 tsp. baking powder
3/4 tsp. salt
1 tsp. cinnamon
1/4 tsp. each ground nutmeg and cloves
1/2 cup oil
3/4 cup honey
2 eggs, lightly beaten
1/2 cup milk

Oil two 8-inch cake pans, and line the bottoms with waxed paper. Heat oven to 350° F.

Sift dry ingredients together. Mix oil with honey, eggs and milk. Add gradually to dry ingredients, mixing well. Pour into prepared cake pans and bake for about 25 minutes, or until a cake tester comes out clean. Cool for 5 minutes. Turn onto a wire rack, peel away the paper and allow to cool.

Quick frosting

4 oz. Neufchatel cheese, at room temperature
1 tbs. lemon juice
3 tbs. honey
1/4 cup soy powder
1/4 tsp. vanilla (optional)

Mix all ingredients together until smooth and well-blended.

Makes enough to frost 2 eight-inch layers.

Bran cookies

1/2 cup vegetable oil
1/2 cup raw sugar
1 egg
1/2 tsp. sea salt
1/2 tsp. cinnamon
1 tbs. nutritional yeast
1 tsp. baking powder
1/2 cup bran flakes
2/3 cup whole wheat flour
1/4 cup soy granules

Heat oven to 350° F. Oil two cooky sheets.

Beat the oil, sugar and egg together until well mixed. Stir in the remaining ingredients in order, mixing after each addition.

Drop the dough on the prepared sheets in half teaspoonfuls. Space cookies about 2 inches apart. Flatten slightly,

Bake 12 to 15 minutes or until lightly browned around edges. Cool on a wire rack.

Baked Indian pudding

1 cup soy powder
3 cups water
1/3 cup cornmeal
1/4 cup soy granules
1/3 cup apple juice
1/2 tsp. sea salt
1/2 tsp. cinnamon
1/2 tsp. powdered ginger
1/4 cup molasses
1/4 cup honey

Heat oven to 325° F. Oil a 3-quart casserole.

Mix the soy powder with the water and heat to boiling. Mix together the cornmeal, granules and apple juice. Slowly add this mixture to the soy powder and water. Cook until slightly thickened, stirring often.

Add all remaining ingredients and mix well. Pour into prepared casserole and bake at 325° F. for about 2 hours, or until it is a rich brown.

Cottage pudding

1/2 cup sesame tahini

1/2 cup honey

2 eggs, lightly beaten

1 cup water

1 cup soy protein isolate

1 cup whole wheat flour

2 tsp. baking powder

1/2 tsp. sea salt

Oil an 8-inch by 8-inch by 2-inch pan. Heat oven to 400°F.

Mix tahini, honey, eggs and water together until smooth. Sift together dry ingredients. Add liquid mixture gradually, mix until smooth. Pour into prepared pan and bake at 400° F. for 20-25 minutes or until top browns lightly. Serve warm topped with warmed applesauce or other fruit sauce.

Serves 6 to 8.

Quick carob cornstarch pudding

2 cups soy milk
1/2 cup carob powder
1/4 cup cornstarch
1 tsp. instant coffee (optional)
3/4 cup water
1/4 cup honey
2 tsp. vanilla

Heat the soy milk in a very large pot until it reaches a boil. Meanwhile, mix the carob, cornstarch and coffee. Stir the water slowly into the carob mixture to make a smooth paste. Add the honey. Slowly add the mixture to the simmering soy milk and stir until thickened. Cook slowly over low heat for about 5 minutes. Let cool for a few minutes, then stir in the vanilla. Spoon into serving dishes and chill.

Serves 6.

Bibliography

Aaron M. Altschul. *Proteins, Their Chemistry and Politics.* New York: Basic Books, 1965.

Carroll and Vona. *The Health Food Dictionary.* Englewood Cliffs, N. J.: Prentice-Hall, 1973.

Philip S. Chen, Ph.D. *Soybeans.* New Canaan, Conn.: Keats Publishing, 1973.

Gayelord Hauser. *Gayelord Hauser's New Treasury of Secrets.* New York: Farrar, Straus & Giroux, 1974.

Johnson and Peterson. *Encyclopedia of Food Technology, Vol. 2.* Westport, Conn.: AVI Publishing, 1974.

Frances Moore Lappé. *Diet for a Small Planet.* New York: Ballantine Books, 1971.

Smith and Circle. *Soybeans: Chemistry and Technology, Vol. 1.* Westport, Conn.: AVI Publishing, 1972.

Watt and Merrill. *Composition of Foods.* Washington: United States Department of Agriculture, 1963.

Wolf and Cowan. *Soybeans as a Food Source.* Cleveland, O.: CRC Press, 1971.

Index of Recipes